Bridge Building

Bridge Designs and How They Work

by Diana Briscoe

Reading Consultant:
Timothy Rasinski, Ph.D.
Professor of Reading Education
Kent State University

Content Consultant:
Walter T. Marlowe, P.E., CAE, M.ASCE
Director, Professional Practices
American Society of Civil Engineers

Red Brick™ Learning

Published by Red Brick™ Learning
7825 Telegraph Road, Bloomington, Minnesota 55438
http://www.redbricklearning.com

Library of Congress Cataloging-in-Publication Data
Briscoe, Diana, 1949–
Bridge building: bridge designs and how they work / by Diana Briscoe.
p. cm.—(High five reading)
Includes bibliographical references and index.
ISBN 0-7368-3853-8 (soft cover)—ISBN 0-7368-3881-3 (hard cover)
1. Bridges—Design and construction—Juvenile literature. 2. Bridges—History—Juvenile literature. I. Title. II. Series.
TG148.B75 2004
624.2—dc22

2004003491

Created by Kent Publishing Services, Inc.
Executive Editor: Robbie Butler
Designed by Signature Design Group, Inc.
Edited by Jerry Ruff, Managing Editor, Red Brick™ Learning
Red Brick™ Learning Editorial Director: Mary Lindeen

Photo Credits:
Pages 4, 10, 30, 35, 48, Bettmann/Corbis; page 7, Paul A. Souders, Corbis; page 12, Chuck Mitchell; pages 14–15, Ruggero Vanni, Corbis; pages 16–17, National Geographic Society; pages 18–19, Ed Bohon, Corbis; pages 20–21, Wild Country/Corbis; page 23, W.A. Sharman, Corbis; page 25, Robert Frerck, Odyssey Productions; pages 29, 31, Jason Hawkes, Corbis; pages 34, 43, Robert Holmes, Corbis; page 37 (top and bottom), Library of Congress; page 38, David Williams, Alamy Images; page 39, E.O. Hoppé, Corbis; page 41, Copley News Service; page 45, Grace Davies, GD Photo; pages 46–47, Angelo Hornak, Corbis; pages 52 (top and bottom), 53, Niagara Falls Thunder Alley/niagarafrontier; page 55, Toby Melville, PA Photos; page 57 (top), Adam Woolfitt, Corbis; page 57 (bottom), Tessa Oksanen, Corbis/Sygma

Printed in the United States of America.

1 2 3 4 5 6 09 08 07 06 05 04

Table of Contents

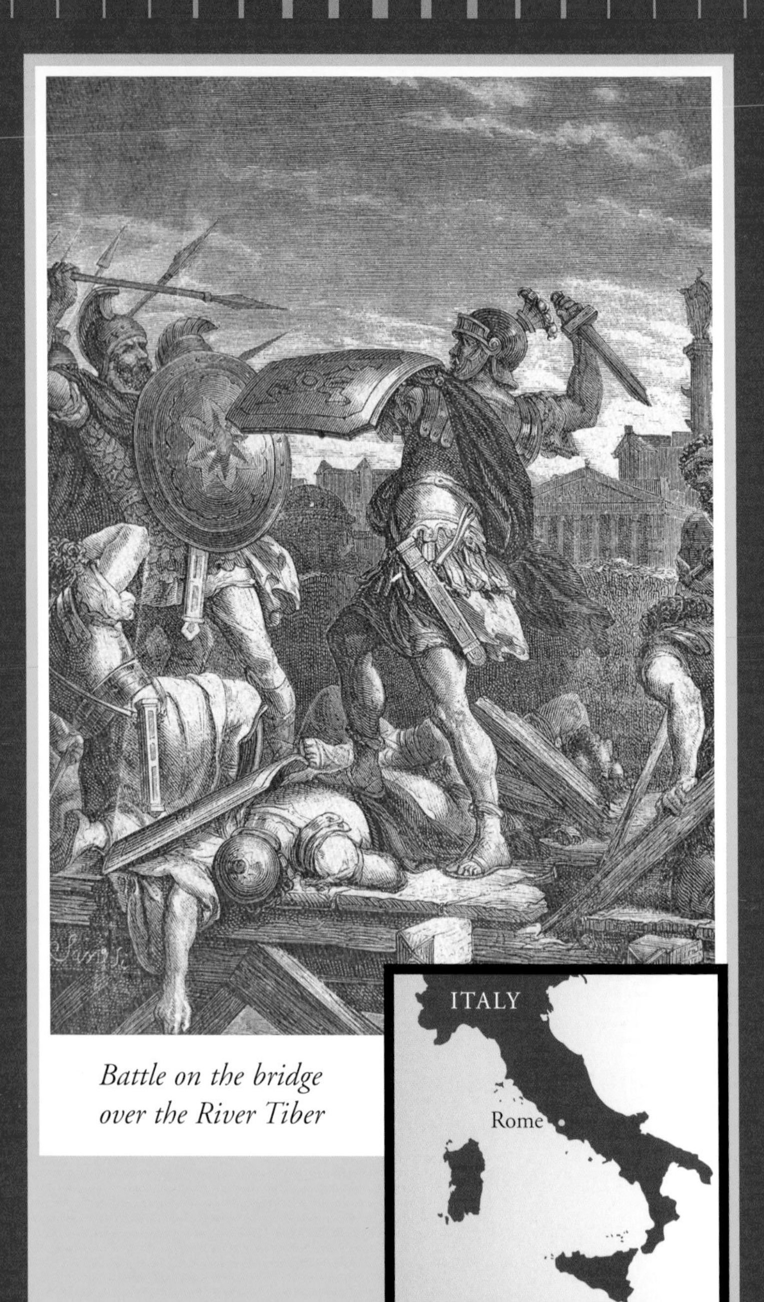

Battle on the bridge over the River Tiber

— Chapter 1 —

The First Bridges

A breathless messenger staggers into the city. "Quick! To the bridge! Tarquin the Proud is coming. He has an army!"

"Find axes!" shouts a Roman soldier.

Tarquin, their ex-king, will show no mercy if he reaches Rome. To stop him, the Romans must destroy the bridge to the city. Roman soldiers race to the River Tiber and start chopping at the bridge.

A True Story

Some 2,500 years ago, a sturdy wooden bridge spanned the River Tiber and led to the city of Rome. Such bridges allowed traffic to come and go easily from the city. Horses, carts, and people could safely cross the fast-flowing river. But so could an enemy, determined to destroy the city and its people.

---

span: to stretch or reach across

What Happened in Rome That Day

The bridge across the River Tiber did not surrender easily to the Roman axes. As Tarquin's army approached, three brave Romans raced across the bridge to hold them back.

At last, the sturdy bridge began to creak and groan. "Back, back, it's going!" someone cried. Two of the men retreated. But the third, named Horatius, continued to fight.

As the bridge began to tumble, Horatius shouted to the river, "Father Tiber, save this Roman's life today!"

The Romans groaned as Horatius fell into the river. But then his head popped up out of the water. He was alive! Cheers erupted as this brave, wet soldier swam back to shore.

---

surrender: to give up to the power of another
erupt: to burst forth suddenly

Historic Tales

Historic battle tales such as this often mention bridges. We have learned some of the history of bridges by reading such tales. The story of Horatius is one of the first to talk about a bridge. It tells us something about how bridges of that time were built.

But people had been building bridges for thousands of years before these Romans. The first bridges were trees or logs placed across streams and ditches. This simplest type of bridge is called a *beam bridge*. A beam bridge is a flat surface resting on two or more supports. The bridge that Horatius defended was a beam bridge.

This log spanning a river is a simple beam bridge.

The Earliest Bridges

More than 4,500 years ago, people in Europe built walkways over rivers and marshes. Some walkways were as long as 1 mile (1.6 kilometers).

Building these walkways took a lot of work. People cut down hundreds of trees. Some of these trees they cut for piers and sunk into the ground. Others were cut for beams to span the piers. Still other trees were split into planks and placed over the beams to make a deck for the walkway. Bridge builders braided miles of rope to tie the planks and piers together. Everybody in the area helped with this huge task.

People still use walkways today. You've probably walked across one of these simple beam bridges yourself!

---

marsh: low land that is wet and soft
pier: an upright support for the middle sections of a bridge
plank: a long, wide, thick board

Early Beam Bridge Construction

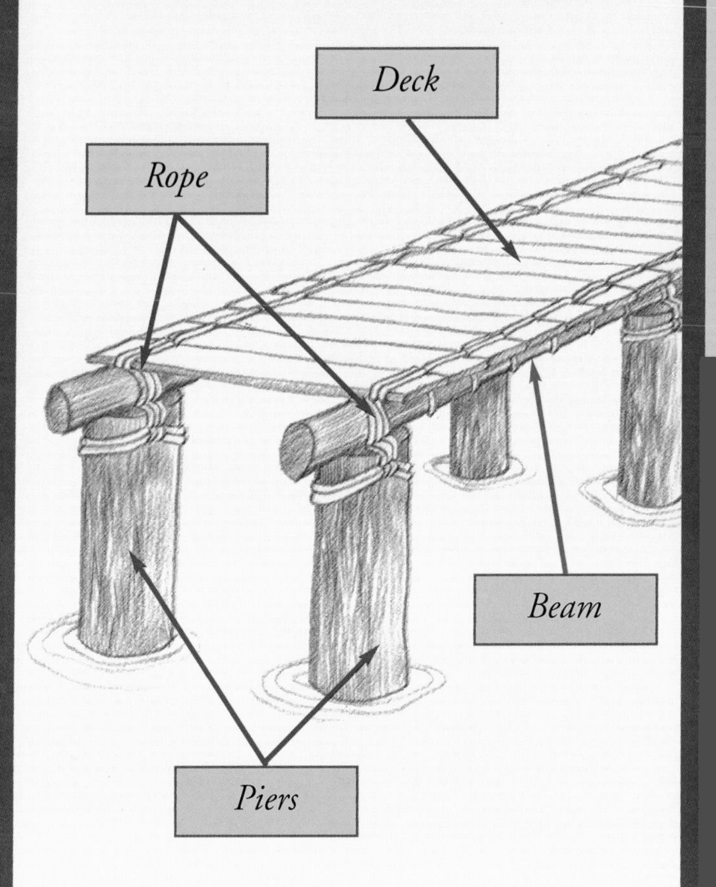

Floating Bridges

Imagine that you want to invade another country, but the sea is in the way! How do you get across? One way is to find the narrowest stretch of water and build a beam bridge across it using boats for support.

This was what Xerxes, the King of Persia, did in 484 B.C. He wanted to invade Greece. But first, he had to move his army across the Hellespont Strait between Asia and Europe. The distance was more than 1 mile (1.6 kilometers). His engineers collected hundreds of boats. Then they tied the boats together and laid a road across the boats.

Xerxes, King of Persia

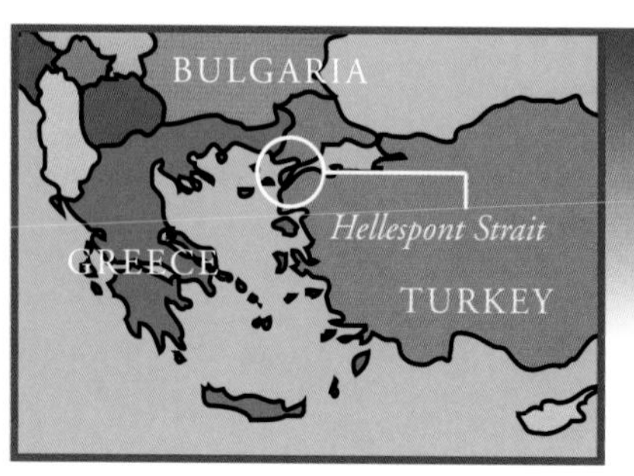

Persia: the country now called Iran
engineer: a person who designs and builds roads, bridges, buildings, and other structures

Conquering the Ocean

Another famous emperor built a bridge of boats, too. In A.D. 39–40, the Roman emperor Caligula used more than 5,000 boats to make a bridge across the Bay of Naples in Italy. Then he rode over the floating bridge and declared he had conquered the ocean!

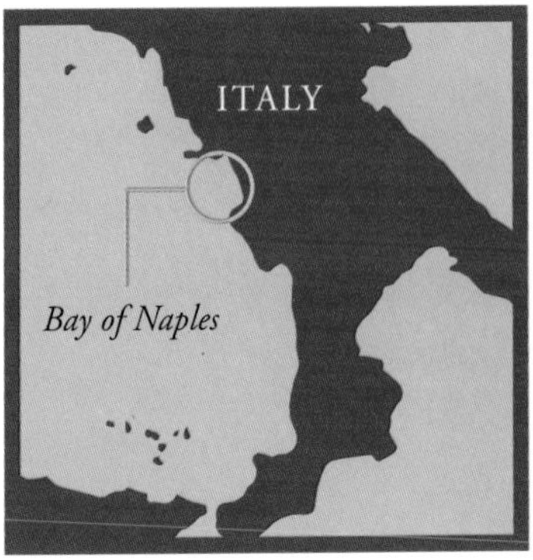

Bridges Begin to Change

Tales of wars have told us something about early bridges. Over time, engineers designed new types of bridges. Why do you think bridge design changed? What do you think these new bridges looked like?

---

conquer: to overcome

— CHAPTER **2** —

Arch Bridges

Bridges let people and other traffic cross over water. Yet traffic isn't only going over bridges, it also goes under them! Boats and ships have to travel under bridges. Long ago, some ships had tall masts. Engineers needed to design a bridge that ships could sail beneath. The arch was the answer.

Pine Creek Bridge in Zion National Park, Utah

The First Arch Builders

The ancient Egyptians built the first arches nearly 5,000 years ago. These arches could not hold heavy weight, though. Around 500 B.C., the Romans learned to build a stronger **semicircular** arch. To build it, they fit stones, called *voussoirs* (voo-SWAHRS), together in a curve. The last stone to go in at the top of the curve they called the *keystone*. The keystone held everything in place.

The Romans used this semicircular arch in their bridges. An arch bridge allowed taller traffic to pass underneath. It was also strong enough to support traffic crossing over.

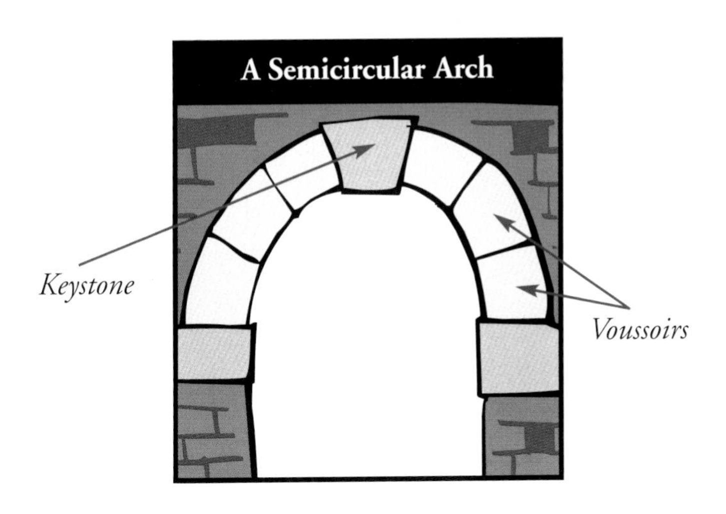

semicircular: in the shape of half a circle

Water Overhead

The Romans didn't build bridges just to carry people over water. They built bridges to carry water over people as well!

Romans used large amounts of water in their towns. To get this water, they built aqueducts (ACK-wuh-duhkts). These arch bridges, or groups of arch bridges, carried water from the hills into towns below.

The Claudian Aqueduct near Rome

Water from the hills flowed into a channel at the top of the aqueduct. Then the aqueduct brought the water downhill into the town.

On its way from hilltop to town, the aqueduct soared over valleys. Some aqueducts even tunneled through cliffs and crossed rivers. Other aqueducts were multi-level bridges with water flowing on the top level and people traveling on the lower level.

Anji Bridge

About A.D. 580, in northern China, engineer Li Chun had a problem. If he built a beam bridge over the fast-flowing Xiao (shee-OW) River, boats could not get underneath. But a semicircular arch bridge over this wide river would be too steep. In fact, at the middle, it would have to be 66 feet (20 meters) high. People could not pull their carts up such a steep slope.

Li Chun's Anji Bridge is still in use after more than 1,400 years.

The spandrel is the triangular area at either end of the bridge between the arch and the deck.

Li Chun had an idea. He would build a longer, flatter bridge. But it would still have enough arch to let boats pass underneath. In other words, the arch would form a segment, or part, of a circle. This design is called a *segmented bridge*.

Li Chun also made the bridge lighter by putting smaller arches in the spandrels. His thinking was way ahead of his time. Other bridge builders would not use this design for several hundred years to come.

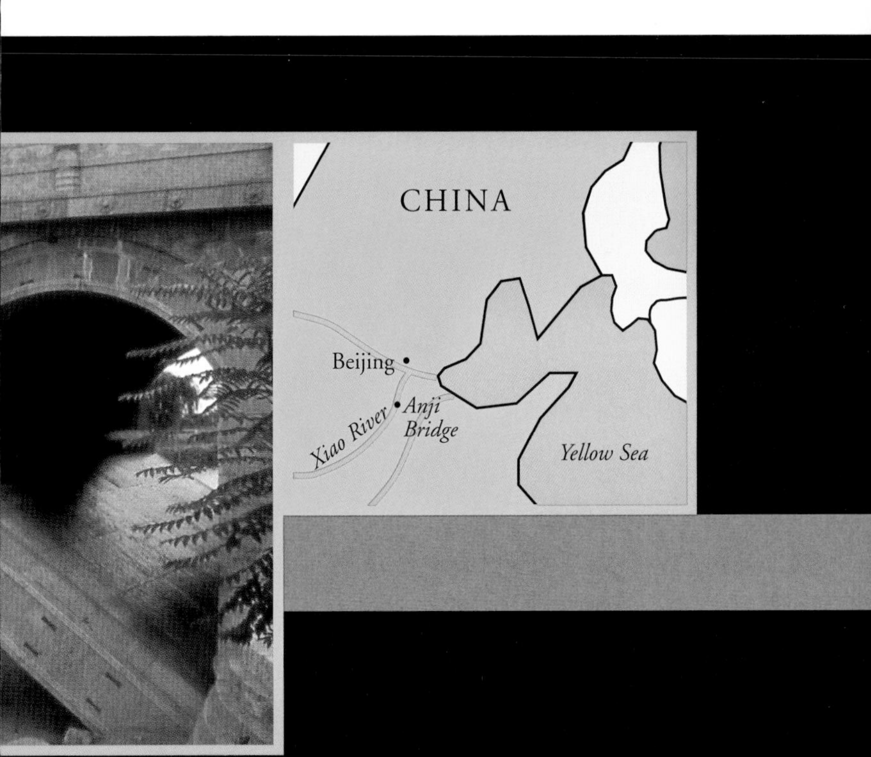

Life on a Bridge

Arch bridges were strong and could hold up under heavy weight. In the Middle Ages, people built houses and shops on bridges. Bridges became neighborhoods and places to buy and sell goods. The only such bridge that remains today is the Ponte Vecchio (PON-tuh VEK-ee-oh) in Florence, Italy.

Middle Ages: a time in European history between A.D. 476 and A.D. 1450

The Ponte Vecchio is also a segmented bridge like the Anji. It was built around 1334 to replace an older bridge destroyed in a flood. An upper level was added to the Ponte Vecchio in 1565.

The Ponte Vecchio spans the Arno River in Florence, Italy.

The Iron Bridge

People could only build bridges from the materials available at the time. Until the 1700s, these materials were wood, stone, and concrete.

But bridge building would soon change. In the 1770s, iron masters from Coalbrookdale, England, discovered how to smelt iron. To do this, they melted iron ore and removed waste matter from it. Smelting made the iron much stronger. It could also be easily molded. Now, iron could be used to build bridges.

The Iron Bridge spans 100 feet (30.5 meters) over the River Severn.

In 1779, engineers in Britain used iron to build a new arch bridge over the River Severn. They cast the Iron Bridge in 800 different sections. Workers put the bridge together in only three months.

The design of the Iron Bridge was not new. But it was the world's first major structure built entirely of iron. Also, it was the world's first large-scale, cast-iron bridge.

The Iron Bridge soon proved its strength. It was the only bridge to survive huge floods in 1795. From then on, many other bridge builders only worked with iron.

cast: to form or shape melted metal
cast iron: hard, brittle iron shaped by casting

Avoiding Ups and Downs

Railroads arrived in the 1800s. To support the weight of trains and tracks, bridge builders used the same design idea used by the Romans who built the aqueducts.

Bridge builders began to build viaducts (VYE-uh-duhkts) over rivers and valleys. Viaducts carried trains on a flat surface so they didn't have to go up and down steep hills. In Europe, most viaducts were arch bridges built of brick or stone.

At first, North American viaducts were built using concrete slabs made with cement. But it took too much time and money to build these bridges. American railroad companies wanted to quickly add more and more railroads. So, they ordered viaducts built of wood—a much cheaper and faster way to build.

---

viaduct: a bridge, or line of bridges, that carries a road or railroad across a valley
slab: a piece of wood, concrete, or stone that is flat, broad, and thick

A steam train crosses a viaduct in Knaresborough, England.

The Outdated Arch

New materials and new designs had changed bridge building greatly since the simple beam bridge. Bridges could now span much longer distances and carry greater loads. Soon, however, even arch bridges would also become outdated.

What new designs for building bridges would come next? Any ideas?

---

outdated: behind the times; no longer current

— CHAPTER **3** —

Suspension Bridges

Have you ever walked across a swinging bridge at a playground or an amusement park? Then you have crossed a suspension bridge. The Golden Gate Bridge and Brooklyn Bridge are suspension bridges, too. Hanging from chains or cables, a modern suspension bridge was something new to people less than 200 years ago. Imagine what it felt like to cross one for the first time!

The Golden Gate Bridge

Jungle Roads

The Incas in South America built the first suspension bridges about 500 years ago. These highly civilized people made bridges using ropes and wood. They braided or twisted fibers, vines, or thin twigs to make strong ropes—some as much as 16 inches (40 centimeters) thick! Each bridge used five such ropes. Three ropes supported the deck, which was made of sticks, matting, and mud. The other two ropes worked as handrails.

The Incas used these rope bridges to cross deep gorges. Some of these bridges were up to 200 feet (61 meters) long.

This bridge is made of rope.

Incas: a people of ancient Peru
matting: a flat piece of rough material, such as straw and twigs, woven together
gorge: a narrow valley between steep cliffs

Hanging from a Wire

Modern suspension bridges took the Incan design and "super-sized" it! The design works like this: Two stone piers are built near either end of the span the bridge will cross. A chain is strung across the top of the piers and then anchored into solid rock at either end. Suspension rods hang from the chain down to the bridge deck. The weight of the deck pulls down on these rods and the chain, but the piers take this force. That way, the bridge doesn't sag or collapse.

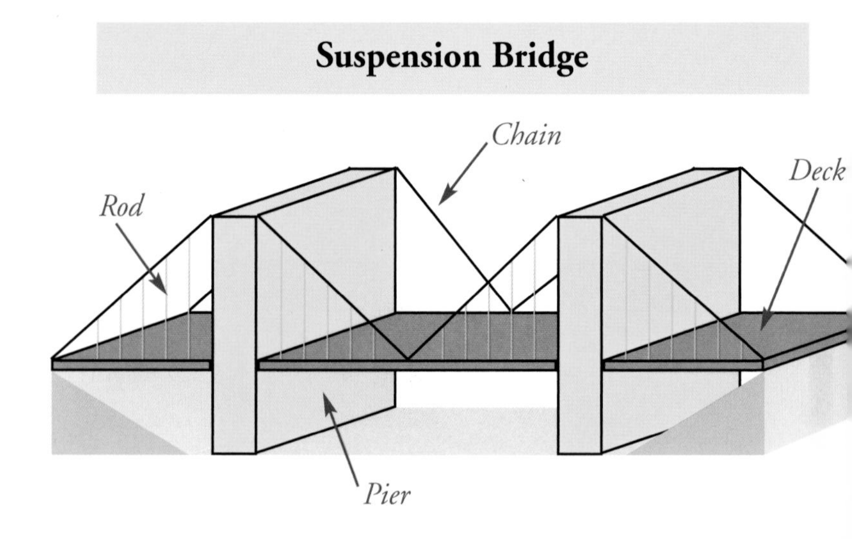

---

span: the distance between two ends
anchor: to secure something so it will not move

Rods hanging from cables hold up the deck on the Verrazano-Narrows Bridge in New York.

Chains from Shore to Shore

Thomas Telford (1757–1834) built 1,200 bridges in his lifetime. But his greatest work is the Menai Suspension Bridge connecting Anglesey and mainland Wales in the United Kingdom.

This early suspension bridge was special because of its strength. Wrought-iron chains carried the load on the Menai Suspension Bridge. Work on the bridge began in 1819. When the bridge was finished in 1826, 16 chains supported the bridge. The main span is 579 feet (176 meters) long. But the total length of the bridge is 1,709 feet (521 meters). Some 444 vertical rods connect the deck to the chains.

---

wrought-iron: made of iron that is hammered into a shape
vertical: straight up and down

The wrought-iron chains of the Menai Suspension Bridge were replaced with steel chains in 1942.

A Leap over the Gorge

The Clifton Suspension Bridge crosses the Clifton Gorge and the River Avon near Bristol, England. Isambard Kingdom Brunel (brew-NELL) designed the bridge in 1830, and work began in 1831.

Isambard Kingdom Brunel

In 1843, the project ran out of money. Brunel died in 1859, with the bridge still not finished. Because Brunel loved this bridge so much, friends completed it in 1864 as a memorial to him.

The bridge was built to support the horse-drawn traffic of its time. Amazingly, today the same bridge supports the traffic of 4 million cars every year.

The Clifton Suspension Bridge

---

memorial: anything meant to remind people of a person or an event

The Brooklyn Bridge

John A. Roebling (ROH-bling) designed the Brooklyn Bridge in 1869. He said he wanted the bridge to be "ranked as a national monument… a great work of art." His design was artistic and innovative.

Roebling was the first to use steel in bridge building. He called steel "the metal of the future," and used this type of iron to make suspension cables. He also developed a new way of attaching the cables on a bridge.

The Brooklyn Bridge spans the East River between Manhattan and Brooklyn, New York.

---

artistic: done with imaginative skill
innovative: a new way of doing something

Roebling died from blood poisoning in 1869, before construction on the bridge began. But his son, Washington A. Roebling, took over to ensure the bridge would be built.

All went well until an accident in 1872 left Washington crippled, nearly blind, and unable to write. His wife, Emily, then took over managing the bridge project. Emily would write out Washington's instructions and explain them to the workers. The bridge was finally completed in 1883.

Not Suited for Everything

Suspension bridges worked well in the 1800s to span huge lengths. But they were flexible, and wind or other forces could make them move. This meant that they were not suited for heavy loads that produced a lot of motion, such as steam trains. So what could be done to carry heavy, moving loads over very long spans? The answer is something you've seen on the playground.

— CHAPTER **4** —

Trestles, Trusses, and Cantilevers

From the 1820s on, bridge builders constructed better and better suspension bridges. At the same time, they tried to improve the older beam bridge design. Three types of beam bridges were developed—trestle, truss, and cantilever. Look at the A-shaped frame at the ends of this swing set. This frame is called a trestle.

Trestle Bridges

For a beam bridge to carry a heavier weight, such as a train, it had to have stronger supports. So engineers designed trestles similar to the supports in a swing set. They spaced the trestles close together to enable the bridge to carry heavier weight.

Because wooden bridges were quicker and cheaper to build than stone bridges, many early trestle bridges that carried trains were made from wood.

Dozens of trestles are linked together to support this rail bridge built in the 1800s.

Truss Bridges

Squire Whipple was an engineer who also studied the problem of how to carry a heavy load over a long bridge. He had an idea that trusses could make these bridges stronger. A truss is a beam made from a series of triangles locked together. Each of the short pieces of the triangle carries a small part of the load. This spreads a heavy load out over more pieces. A jungle gym is made of trusses.

This jungle gym uses trusses to support the weight of children playing on it.

Whipple's Idea

Squire Whipple

In 1841, Whipple learned that many bridges had been built using no mathematical or scientific study about the stress on a bridge. So Whipple studied what happened to a bridge under stress. He used what he learned to design a type of truss bridge. One of his designs is known as the Whipple Bowstring Truss Bridge.

Squire Whipple's Bowstring Truss Bridge near Utica, New York

---

stress: strain, force, or pressure

Cantilever Bridges

What happens when someone bigger than you sits on one end of a seesaw? You go up in the air and stay there! When you put more weight at one end of the seesaw, the other end goes up. If the weight is equal, the seesaw stays level. That's how a cantilever works.

Imagine two seesaws lined up. Each has equal weight at both ends. Now, connect the ends where the two seesaws meet. You've made a cantilever bridge!

Both of these children are about the same weight, so the seesaw stays nearly level.

We Need Steel!

The first cantilever bridges were built in China and Tibet more than a thousand years ago. But they were made of wood and could not carry anything heavier than a herd of goats.

It was not until steel was made in the 1870s that the cantilever design could be used in modern bridges. The Firth of Forth Bridge built in Scotland is an example of a steel cantilever bridge.

The Firth of Forth Bridge is made like seesaws connected together.

Build toward the Middle

Today, many bridge designs combine several methods of bridge building. For example, an arch bridge may also use cantilevers and suspension cables. Here's how.

Let's start with the cantilevers. These are used at both banks of a river or harbor as workers start to build the arch from both ends, moving toward the middle. To do this, the workers dig huge tunnels into the rock on each bank. They anchor cables into these tunnels. The cables will then support the two sides of the arch as it is being built.

Two cranes are used to build the arch. A crane starts at each end, then moves up little by little, building the arch as it goes! To build the arch, each crane hoists steel up from barges below and fixes it into place on the growing arch. When the two cranes meet at the top, the arch is complete. Workers then build the deck, hanging it from suspension cables from the arch.

---

hoist: to lift or pull up
barge: a large boat with a flat bottom used to carry goods on rivers and canals

The Sydney Harbour Bridge

The Sydney Harbour Bridge in New South Wales, Australia, is an arch bridge that uses this design. To build it, workers used cantilevers at each bank and worked toward the middle or top of the arch. When the arch was complete, they hung steel cables from the arch to support the deck.

During construction, a gale hit the Sydney Harbour Bridge just before the two halves of the arch met. Yet the high winds did no damage at all to this very strong bridge. The bridge was finished in 1932.

The Sydney Harbour Bridge in Australia spans 1,650 feet (503 meters).

---

gale: a strong wind

Longer and Longer

Today, bridges, or a series of bridges, cover longer and longer distances. The longest combination of bridges in the world is in Louisiana. The Interstate 55 and Interstate 10 highways are carried on twin concrete trestles for more than 34 miles (55 kilometers) near Manchac (MAN-shack).

Two other very long causeways are also in Louisiana. They are the Lake Pontchartrain (PONT-shur-trayn) I and II Bridges. The Lake Pontchartrain II Bridge is the world's longest road bridge at 24 miles (39 kilometers) long. The two bridges run from Metairie (MET-uh-ree) to Lewisburg, over Lake Pontchartrain.

---

causeway: a raised road across a marsh or stretch of water

Each weekday, more than 30,000 cars cross over the Lake Pontchartrain causeway.

Moving Bridges

In some places, it is impossible to build a bridge high enough for ships to pass underneath. In this case, you need a moving bridge that can open to let a ship through. Three types of moving bridges are swing, bascule, and double bascule bridges. All three are cantilever bridges.

Swing Bridges

There are many ways to make a bridge that opens. The most important thing to know is how much weight the bridge has to carry. A swing bridge is often used if the bridge only has to carry a light weight. Swing bridges have a huge pivot underneath. The whole bridge turns on the pivot to let a ship through.

---

pivot: a point, pin, or rod upon which something turns

This swing bridge in Jamaica Bay, New York, is halfway open.

Bascule Bridges

When trucks and heavy traffic must be able to cross, a bascule bridge is used. One side of the bridge has a very heavy counterweight on it. The bridge tips up into the air when the locks that hold the weight are released.

On long spans, both sides of the bridge have a counterweight and tip up. This is called a double bascule. The Tower Bridge in London, England, is a famous double bascule bridge.

The Tower Bridge in London, England

counterweight: a weight at one end of a beam

Always Strong and Safe?

Bridges must be strong and safe. A well-built bridge will usually stand for many years without problems. But sometimes, things don't go so well. What do you think might cause a bridge to fail? Read the next chapter to find out.

— CHAPTER **5** —

Great Bridge Disasters

Most bridges are safe, strong, and even beautiful. But once in a while, a bridge will fail. Sometimes, it is bad design that causes the problem. Other times, an accident may be to blame. Whatever the reason, when a bridge fails, it is a very serious event.

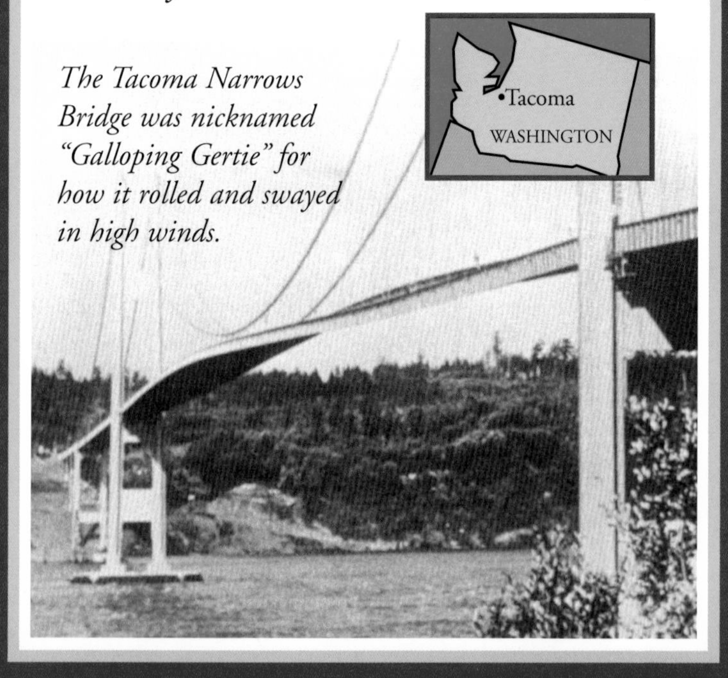

The Tacoma Narrows Bridge was nicknamed "Galloping Gertie" for how it rolled and swayed in high winds.

The Tacoma Narrows Bridge

The original Tacoma Narrows Bridge in Washington opened to traffic on July 1, 1940. The $6.4-million bridge linked Tacoma and Gig Harbor over Puget (PEW-jet) Sound.

The Tacoma Narrows Bridge also had a nickname—"Galloping Gertie." It got this nickname for the way the bridge would sway and roll in high winds.

Drivers on the 2,800-foot (853-meter) center span compared it to riding on a giant roller coaster. Some said it made them seasick. When the rolling was at its worst, the sidewalk on one side of the bridge was 28 feet (8.5 meters) higher than the sidewalk on the other side!

Tumbling Down

On November 7, 1940, the wind was blowing at 42 miles (68 kilometers) per hour. Galloping Gertie rolled as usual, but this time, that wasn't all she did. First, a piece of concrete fell into the water below. But that was only the beginning. The wind and the twisting continued, until finally a 600-foot (183-meter) section of the deck broke off. The section turned upside down as it crashed into Puget Sound.

Leonard Coatsworth, a Tacoma newspaper editor, was on the bridge that day. He wrote this account:

"Just as I drove past the towers, the bridge began to sway violently from side to side. Before I realized it, the tilt became so violent that I lost control of the car.... I jammed on the brakes and got out, only to be thrown onto my face against the curb.

"Around me I could hear concrete cracking. I started to get my dog, Tubby, but was thrown again before I could reach the car. The car itself began to slide from side to side of the roadway.

"On hands and knees most of the time, I crawled 500 yards or more to the towers.... My breath was coming in gasps; my knees were raw and bleeding, my hands bruised and swollen from gripping the concrete curb.... Toward the last, I risked rising to my feet and began running a few yards at a time.... Safely back at the toll plaza, I saw the bridge in its final collapse and saw my car plunge into the Narrows."

Amazingly, no people died in this disaster. Coatsworth's dog, Tubby, however, did die when the bridge collapsed.

The Tacoma Narrows Bridge isn't the only bridge to fail. Read on to learn more about another bridge disaster.

The Falls View Bridge, Niagara Falls

This steel arch bridge was completed in 1898. One year later, enormous chunks of ice threatened to crush it. Some chunks were 80 feet (24 meters) high, reaching to the base of the arch. The pressure from the ice pushing on it was so great that several pieces of the steel arch bent. The next summer, workers built extra walls to hold the winter ice back and protect the bridge.

The Falls View Bridge

The Falls View Bridge after its collapse

These extra walls worked well for nearly 40 years. Then on January 23, 1938, a sudden windstorm jammed ice in the river below the Falls. As this ice pushed against the bridge, it caused a lot of damage.

The bridge held on for four days. Thousands came to see it. But finally, at 4:20 P.M. on January 27, the bridge went crashing onto the ice below. Two pieces of the center span remained on the ice until April 12, when the ice finally melted and the pieces sank.

Remains of the Falls View Bridge are still there today.

Not Supposed to Sway

The Millennium Bridge is a footbridge over the Thames River in London, England. This suspension bridge has no towers and runs 1,072 feet (327 meters) long.

The Millennium Bridge opened on June 10, 2000. That day, more than 80,000 people crossed it. As large groups of people walked across the Millennium Bridge, it began to sway. Two days later, the bridge was closed so engineers could study this problem.

The engineers found that the bridge swayed because of how people transfer their weight from side to side when walking. As the bridge swayed, people would sway with it. This caused the bridge to sway more and more.

---

transfer: to move, carry, send, or change from one person or place to another

The Millennium Bridge

Problem Solved

To solve the problem, engineers put dampers under the deck and around the piers. The dampers acted like shock absorbers in cars. They absorbed the movement that was being transferred from the people to the bridge. The solution worked perfectly, and the bridge reopened in August 2002.

Come a Long Way

Bridge design has come a long way from its early wooden-beam beginnings. Today, bridge designers use computer programs to test how strong a bridge will be before it is even built.

Technology like this, along with improved construction materials, makes today's bridges the safest in history. Some people think these bridges are the most beautiful, too.

What is your favorite bridge design? Can you invent a bridge design of your own?

An ancient Roman aqueduct and the modern Millennium Bridge

Epilogue

New Bay Bridge

On October 17, 1989, the Loma Prieta (LOH-mah pree-YAY-tah) earthquake began. The tremors damaged the eastern span of the San Francisco-Oakland Bay Bridge. The 7.1 quake knocked down part of the bridge's upper deck, killing one motorist. The bridge was closed for a month for repairs.

Engineers studied the damaged bridge. They decided it would be safer and cheaper to build a new bridge rather than to try to "earthquake-proof" the old one.

The new east span is now being built. The span has four main parts:

- Oakland Touchdown—A short, low-rise span that connects the Oakland shore to the bridge .
- Skyway—A 1.5-mile-long roadway from the low-rise span to the suspension section.
- Suspension Section—The world's first single-tower, self-anchored suspension span.
- Yerba Buena (YAIR-bah BWAY-nah) Transition—A roadway that connects the suspension section to the Yerba Buena tunnel.

Engineers hope to finish the east span by 2007. But this new bridge won't be cheap. It will cost $2.6 billion! To learn more, check out this Internet site: http://newbaybridge.org/.

Glossary

anchor: to secure something so it will not move

artistic: done with imaginative skill

barge: a large boat with a flat bottom used to carry goods on rivers and canals

cast: to form or shape melted metal

cast iron: hard, brittle iron shaped by casting

causeway: a raised road across a marsh or stretch of water

conquer: to overcome

counterweight: a weight at one end of a beam

engineer: a person who designs and builds roads, bridges, buildings, and other structures

erupt: to burst forth suddenly

gale: a strong wind

gorge: a narrow valley between steep cliffs

hoist: to lift or pull up

Incas: a people of ancient Peru

innovative: a new way of doing something

marsh: low land that is wet and soft

matting: a flat piece of rough material, such as straw and twigs, woven together

memorial: anything meant to remind people of a person or an event

Middle Ages: a time in European history between A.D. 476 and A.D. 1450

outdated: behind the times; no longer current

Persia: the country now called Iran

pier: an upright support for the middle sections of a bridge

pivot: a point, pin, or rod upon which something turns

plank: a long, wide, thick board

semicircular: in the shape of half a circle

slab: a piece of wood, concrete, or stone that is flat, broad, and thick

span: (verb) to stretch or reach across; (noun) the distance between two ends

stress: strain, force, or pressure

surrender: to give up to the power of another

transfer: to move, carry, send, or change from one person or place to another

vertical: straight up and down

viaduct: a bridge, or line of bridges, that carries a road or railroad across a valley

wrought-iron: made of iron that is hammered into a shape

Bibliography

Adkins, Jan. *Bridges: From My Side to Yours.* Brookfield, Conn.: Roaring Brook Press, 2002.

Brown, David. *Bridges.* New York: Macmillan Publishing Company, 1993.

Curlee, Lynn. *The Brooklyn Bridge.* New York: Atheneum Books for Young Readers, 2001.

Johmann, Carol A. and Elizabeth J. Reith. *Bridges! Amazing Structures to Design, Build & Test.* A Kaleidoscope Kids Book. Charlotte, Vt.: Williamson, 1999.

Useful Addresses

American Society of Civil Engineers
1801 Alexander Bell Drive
Reston, VA 20191

National Society of Professional Engineers
1420 King Street
Alexandria, VA 22314-2794

Society of Women Engineers
230 E. Ohio Street, Suite 400
Chicago, IL 60611-3265

Internet Sites

American Society of Civil Engineers—
Kids and Careers
http://www.asce.org/kids/

Bridge Building
http://42explore.com/bridge.htm

Building Big Bridges
http://www.pbs.org/wgbh/buildingbig/bridge/

Index